Long Division
A complete workbook with lessons and problems

By Maria Miller

Contents

Preface

Hello! I am Maria Miller, the author of this math book. I love math, and I also love teaching. I hope that I can help you to love math also!

I was born in Finland, where I also grew up and received all of my education, including a Master's degree in mathematics. After I left Finland, I started tutoring some home-schooled children in mathematics. That was what sparked me to start writing math books in 2002, and I have kept on going ever since.

In my spare time, I enjoy swimming, bicycling, playing the piano, reading, and helping out with Inspire4.com website. You can learn more about me and about my other books at the website MathMammoth.com.

This book, along with all of my books, focuses on the conceptual side of math... also called the "why" of math. It is a part of a series of workbooks that covers all math concepts and topics for grades 1-7. Each book contains both instruction and exercises, so is actually better termed *worktext* (a textbook and workbook combined).

My lower level books (approximately grades 1-5) explain a lot of mental math strategies, which help build number sense — proven in studies to predict a student's further success in algebra.

All of the books employ visual models and exercises based on visual models, which, again, help you comprehend the "why" of math. The "how" of math, or procedures and algorithms, are not forgotten either. In these books, you will find plenty of varying exercises which will help you look at the ideas of math from several different angles.

I hope you will enjoy learning math with me!

Introduction

Long Division Workbook includes lessons on long division and remainder. It is suitable for students in fourth grade.

In the lesson *The Remainder, Part 1*, we study the concept of remainder, first using pictures and small numbers. In the second lesson on remainder, we still use small numbers, but students work the problems using the long division symbol or "corner", as I like to call it. That is of course preparing them for long division.

Next, long division is taught in several small steps over many lessons. We start with the situation where each of the thousands, hundreds, tens, and ones can be divided evenly by the divisor. Then is introduced the remainder in the ones. Next comes the situation where we have a remainder in the tens; then we have a remainder in the hundreds, and so on. We also have lots of word problems to solve.

I wish you success in teaching math!

Maria Miller, the author

Helpful Resources on the Internet

MathFrog Dividerama!
Interactive long division practice. Guided help available.
http://cemc2.math.uwaterloo.ca/mathfrog/english/kidz/div5.shtml

Mr. Martini's Classroom: Long Division
An interactive long division tool.
http://www.thegreatmartinicompany.com/longarithmetic/longdivision.html

Drag-and-Drop Math
Practice division interactively. Choose "Division", 2-digit dividend, and 1-digit divisor.
http://mrnussbaum.com/drag-and-drop-math/

Long Division Millionaire Game
Learn to divide large numbers up to thousands. Can you answer all 15 questions?
http://www.kidsmathtv.com/free/long-division-game-for-6th-grade-millionaire-game/

Division Jump — board game
Practice division of one-digit numbers into two, three, and four-digit numbers.
http://www.learn-with-math-games.com/division-activities.html

Bike Racing Math Average
Race your motorcycle against others while answering questions about average. Speed up with each correct answer!
http://www.mathnook.com/math/bike-racing-math-average.html

Long Division Quiz
Practice dividing four-digit numbers by single-digit numbers in this online quiz.
http://i4c.xyz/nmenbdv

Double-Division.org
Double-division is a form of the long division algorithm that takes away the guesswork of finding how many times the divisor goes into the number to be divided. Also called 1-2-4-8 division.
http://www.doubledivision.org/

Short Division
This is a web page that explains short division in detail. Short division is the same algorithm as long division, but some steps are only done in your head and not written down.
http://www.themathpage.com/ARITH/divide-whole-numbers.htm

The Remainder, Part 1

Sometimes we can't divide objects into groups evenly and some of the objects are left over. Those "leftovers" are the **remainder**. We mark the remainder in division with the letter R.

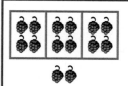 Here you see 14 berries divided into 3 groups as evenly as possible.

We can write the division $14 \div 3 = 4 \text{ R2}$. The divisor (3) tells us how many groups we made. The answer (4) tells us how many berries are in each group. The remainder is 2 berries.

1. Divide the objects into as many groups as indicated. Write a division. There will be a remainder.

a. Divide into three groups.	**b.** Divide into five groups.	**c.** Divide into four groups.

____ ÷ _3_ = ____ R __	____ ÷ ____ = ____ R __	____ ÷ ____ = ____ R __

2. Write a division with a remainder to match the picture. The number of groups gives you the divisor.

a.	**b.**	**c.**
	 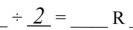	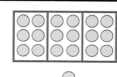
____ ÷ _3_ = ____ R ___	____ ÷ _2_ = ____ R ___	____ ÷ ____ = ___ R ___
d.	**e.**	**f.**
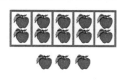		
____ ÷ ____ = ____ R ___	____ ÷ ____ = ___ R ___	____ ÷ ____ = ___ R ___

3. Draw a picture to match the division problem, and solve.

a. $17 \div 4 =$ ____ R ____	**b.** $9 \div 2 =$ ____ R ____	**c.** $11 \div 6 =$ ____ R ____

Besides dividing objects into so many groups, we can also divide them into *groups of a certain size*.

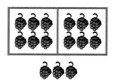 These 15 berries are divided into <u>groups of 6</u>. How many groups do we get?

We can write the division **15 ÷ 6 = 2 R3** . This time the divisor (6) tells us how many berries there are in each group. The answer (2) tells us how many groups we got. Then there is also a remainder of 3 berries.

4. Divide the things into groups of a certain size. Write a division. There will be a remainder.

a. Divide into groups of 4.	**b.** Divide into groups of 2.	**c.** Divide into groups of 5.

_____ ÷ _____ = _____ R __	_____ ÷ _____ = _____ R __	_____ ÷ _____ = _____ R __

5. Draw a picture to match the division problem, and solve. Think of making groups of a certain size.

a. Divide 16 into groups of 5.	**b.** Divide 17 into groups of 3.	**c.** Divide 15 into groups of 4.
_____ ÷ _____ = _____ R __	_____ ÷ _____ = _____ R __	_____ ÷ _____ = _____ R __

Find the remainder by thinking of the DIFFERENCE.

Example. What is 35 ÷ 6?

Think how many groups of 6 there are in 35, or how many times 6 goes into 35.
You can find out with multiplication: 5 × 6 = 30; 6 × 6 = 36. So, 6 goes into 35 five times.

Now find the difference between (5 × 6) and 35, or in other words between 30 and 35.
That difference is 5, and it is the remainder. So 35 ÷ 6 = 5 R5.

6. Solve.

a. 27 ÷ 5 = _____ R ___ How many times does 5 go into 27?	**b.** 16 ÷ 6 = _____ R ___ How many times does 6 go into 16?	**c.** 11 ÷ 2 = _____ R ___ How many times does 2 go into 11?
d. 37 ÷ 5 = _____ R ___	**e.** 26 ÷ 3 = _____ R ___	**f.** 56 ÷ 9 = _____ R ___
g. 43 ÷ 5 = _____ R ___	**h.** 34 ÷ 6 = _____ R ___	**i.** 40 ÷ 7 = _____ R ___

7. Solve.

a.	b.	c.
23 ÷ 4 = _____ R _____	16 ÷ 7 = _____ R _____	21 ÷ 8 = _____ R _____
23 ÷ 5 = _____ R _____	20 ÷ 3 = _____ R _____	12 ÷ 9 = _____ R _____

8. Divide and find the remainder. Notice the patterns!

a.	b.	c.
10 ÷ 5 = _2_ R _0_	17 ÷ 3 = _____ R ___	12 ÷ 4 = _____ R ___
11 ÷ 5 = _____ R ___	18 ÷ 3 = _____ R ___	13 ÷ 4 = _____ R ___
12 ÷ 5 = _____ R ___	19 ÷ 3 = _____ R ___	14 ÷ 4 = _____ R ___
13 ÷ 5 = _____ R ___	20 ÷ 3 = _____ R ___	15 ÷ 4 = _____ R ___
14 ÷ 5 = _____ R ___	21 ÷ 3 = _____ R ___	16 ÷ 4 = _____ R ___
15 ÷ 5 = _____ R ___	22 ÷ 3 = _____ R ___	17 ÷ 4 = _____ R ___

9. Write a number sentence for each word problem. Indicate the remainder, if any.

a. Jim arranged 27 toy cars into rows of 5. How many rows did he get? Were any left over? 	**b.** The teacher put 19 children into groups of 5. How many groups of 5 did she get? What can she do with the "remainder" children? 
c. Mom baked three dozen cookies. She ate three of them, and put the rest into bags, 6 cookies in each bag. How many bags did she get? _____	**d.** Jerry packaged his 51 magazines in 8 bags. Was he able to do so evenly (the same number of magazines in each bag)?
e. Susan wants to organize 35 chairs into nice even rows. Can she organize them into rows of four chairs? Of five? Of six? Of seven?	**f.** Amy put 38 photographs into a photo album. On each page she could fit six photos. How many photos were on the last page? How many pages were full?

The Remainder, part 2

Division can also be written this way. The answer goes on top of the line.	This is $45 \div 9 = 5$. $$\begin{array}{r} 5 \\ 9\overline{)4\ 5} \end{array}$$	This is $21 \div 3$. Write the answer in the right place. $$3\overline{)2\ 1}$$

1. Divide.

a. $8\overline{)2\ 4}$
b. $5\overline{)4\ 5}$
c. $7\overline{)4\ 9}$
d. $8\overline{)7\ 2}$

You can also find the remainder **by subtracting**. Remember, it is the **difference**—the 'leftovers'.	$$\begin{array}{r} 7 \\ 5\overline{)3\ 6} \end{array}$$ How many times does 5 go into 36? Write the answer on top of the line.	$$\begin{array}{r} 7 \\ 5\overline{)3\ 6} \\ -3\ 5 \\ \hline 1 \end{array}$$ Now multiply $7 \times 5 = 35$. Write 35 under 36. Subtract. You get 1. It is the remainder.

2. Divide and find the remainder by subtracting!

a. $$\begin{array}{r} 6 \\ 5\overline{)3\ 2} \\ -3\ 0 \\ \hline \end{array}$$

b. $$\begin{array}{r} \\ 5\overline{)4\ 4} \\ -4\ 0 \\ \hline \end{array}$$

c. $6\overline{)3\ 7}$

d. $7\overline{)2\ 9}$

e. $8\overline{)4\ 6}$

f. $9\overline{)5\ 2}$

g. $4\overline{)3\ 5}$

h. $9\overline{)5\ 7}$

> To check your answer to a division problem with a remainder, multiply your answer by the divisor, then add the remainder. You should get the number you were dividing.
>
> **Example.** Is the division $67 \div 8 = 8$ R5 correct? Check: $8 \times 8 + 5 = 69$. No, it is not.

3. Check your answers to the divisions in problem #2.

Jane packaged 27 cookies into small containers. Four cookies fit into one container.
How many containers did she need?

You can divide $27 \div 4 = 6$ R3 . Is the answer 6 containers, and 3 cookies left outside?
If she puts the 3 cookies into a container too, she will actually need 7 containers!
But if she decides to eat or give away the 3 "remainder" cookies, then she only needs 6 containers.

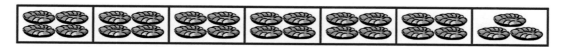

4. Jill put 33 cookies into containers. Six cookies fit into each container.
   How many containers did she need?

   How many full containers did she get?

5. One hundred school children traveled to a pool in buses. Each bus could fit 42 children.
   How many buses were needed?

6. The gym leader divided 20 players into three teams, as evenly as possible.
   How many were in each team?

7. Jessica printed 73 pages of worksheets and put them into folders.
   Each folder could fit 20 pages. How many folders did she need?

   How many folders were full?

8. Twenty-three children participated in a race. Their team leader gave each of
   them three stickers, and after that, he had 15 stickers left. How many stickers
   did the team leader have originally?

9. The teacher had 36 pencils. She divided them evenly among 11 students and put the
   rest of the pencils back in the cabinet. How many pencils were put in the cabinet?

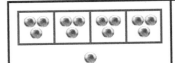	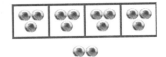	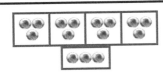
$13 \div 3 = 4$ R1	$14 \div 3 = 4$ R2	$15 \div 3 = 5$ R0
13 divided into groups of 3 makes 4 groups. One is left over.	Add one more marble. It is part of the leftovers.	Add one more. Now, instead of three "leftover" marbles, we can make one more group of 3!

10. Draw one more marble to each picture. Then check if you can make one more group or not. Then write a division sentence

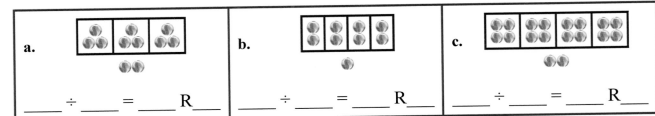

a. _____ ÷ _____ = _____ R_____

b. _____ ÷ _____ = _____ R_____

c. _____ ÷ _____ = _____ R_____

11. Solve, and find a pattern.

a. $21 \div 5 =$ _____ R _____	**b.** $56 \div 8 =$ _____ R _____	**c.** $43 \div 7 =$ _____ R _____
$22 \div 5 =$ _____ R _____	$57 \div 8 =$ _____ R _____	$44 \div 7 =$ _____ R _____
$23 \div 5 =$ _____ R _____	$58 \div 8 =$ _____ R _____	$45 \div 7 =$ _____ R _____
$24 \div 5 =$ _____ R _____	$59 \div 8 =$ _____ R _____	$46 \div 7 =$ _____ R _____

12. Divide by 10. Indicate the remainder. Can you figure out a shortcut?

a. $29 \div 10 =$ _____ R _____	**b.** $78 \div 10 =$ _____ R _____	**c.** $54 \div 10 =$ _____ R _____
$30 \div 10 =$ _____ R _____	$79 \div 10 =$ _____ R _____	$55 \div 10 =$ _____ R _____
$31 \div 10 =$ _____ R _____	$80 \div 10 =$ _____ R _____	$56 \div 10 =$ _____ R _____

Puzzle Corner The number sentence that *checks* the division is given. Write the corresponding division sentence.

a. $5 \times 3 + 1 = 16$	**b.** $7 \times 4 + 3 = 31$	**c.** $4 \times 30 + 15 = 135$
_____ ÷ _____ = _____	_____ ÷ _____ = _____	_____ ÷ _____ = _____

Long Division 1

Divide hundreds, tens, and ones separately.

Write the dividend inside the long division "corner", and the quotient on top.

64 ÷ 2 = ?	**282 ÷ 2 = ?**
Divide tens and ones separately:	2 hundreds ÷ 2 = 1 hundred (h)
6 tens ÷ 2 = 3 tens (t)	8 tens ÷ 2 = 4 tens (t)
4 ones ÷ 2 = 2 ones (o)	2 ÷ 2 = 1. (o)

```
        t o
        3 2
    2 ) 6 4
```

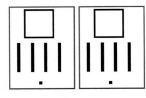

```
      h  t  o
      1  4  1
    2 ) 2  8  2
```

1. Make groups. Divide. Write the dividend inside the "corner" if it is missing.

a. Make 2 groups	**b.** Make 3 groups	**c.** Make 3 groups	**d.** Make 4 groups
2) 6 2	3)	3)	4)

2. Divide thousands, hundreds, tens, and ones separately.

a. 4) 8 4   b. 3) 3 9 3   c. 3) 6 6 0   d. 4) 8 0 4 0

e. 3) 6 6   f. 6) 6 0 3 6   g. 3) 3 3 0   h. 4) 4 8 0 4

h t o 　　0 4)2 4 8	h t o 　0 6 2 4)2 4 8

4 does not go into 2. You can put zero in the quotient in the hundreds place or omit it. But 4 does go into 24, six times. Put 6 in the quotient.

th h t o 　　　0 5)3 5 0 5	th h t o 　0 7 0 1 5)3 5 0 5

5 does not go into 3. You can put zero in the quotient. But 5 does go into 35, seven times.

Explanation:

The 2 of 248 is of course 200 in reality. If you divided 200 by 4, the result would be less than 100, so that is why the quotient won't have any whole hundreds.

But then you combine the 2 hundreds with the 4 tens. That makes 24 tens, and you CAN divide 24 tens by 4. The result 6 tens goes as part of the quotient.

Check the final answer: 4 × 62 = 248.

Explanation:

3,000 ÷ 5 will not give any whole thousands to the quotient because the answer is less than 1,000.

But 3 thousands and 5 hundreds make 35 hundreds together. You can divide 3,500 ÷ 5 = 700, and place 7 as part of the quotient in the hundreds place.

Check the final answer: 5 × 701 = 3,505.

If the divisor does not "go into" the first digit of the dividend, look at the <u>first two digits</u> of the dividend.

3. Divide. Check your answer by multiplying the quotient and the divisor.

a. 3)1 2 3

b. 4)2 8 4

c. 6)3 6 0

d. 8)2 4 8

e. 2)1 8 4

f. 7)4 2 7

g.
　　0 6
3)1 8 3 3

h. 4)2 4 0 4

i. 7)4 9 7 0

j. 5)4 5 0 5

Ones division is not even. There is a remainder.

395 ÷ 3 = 131 R2

```
        h  t  o
        1  3
     3 ) 3  9  5
```

3 goes into 3 one time.
3 goes into 9 three times.

```
        h  t  o
        1  3  1 R2
     3 ) 3  9  5
```

3 goes into 5 one time, but not evenly.
Write the remainder 2 after the quotient.

```
        h  t  o
        0  4  1 R1
     4 ) 1  6  5
```

4 does not go into 1 (hundred). So combine the 1 hundred with the 6 tens (160).

4 goes into 16 four times.

4 goes into 5 once, leaving a remainder of 1.

```
       th  h  t  o
        0  4  0  0 R7
     8 ) 3  2  0  7
```

8 does not go into 3 of the thousands. So combine the 3 thousands with the 2 hundreds (3,200).

8 goes into 32 four times (3,200 ÷ 8 = 400)
8 goes into 0 zero times (tens).
8 goes into 7 zero times, and leaves a remainder of 7.

4. Divide into groups. Find the remainder.

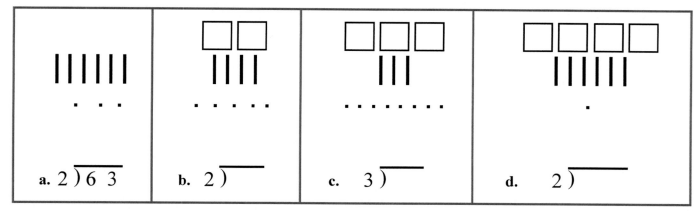

a. 2) 6 3   b. 2)   c. 3)   d. 2)

5. Divide. Indicate the remainder if any.

a. 4) 8 4 7   b. 2) 6 9   c. 3) 3 6 7   d. 4) 8 9

e. 2) 1 2 1   f. 6) 1 8 0 5   g. 7) 2 1 5   h. 8) 2 4 8 2

In the problems before, you just wrote down the remainder of the ones. Usually, we write down the subtraction that actually finds the remainder. Look carefully:

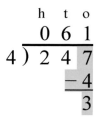

When dividing the ones, 4 goes into 7 one time. Multiply 1 × 4 = 4, write that four under the 7, and subtract. This finds us the remainder of 3.

Check: 4 × 61 + 3 = 247

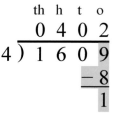

When dividing the ones, 4 goes into 9 two times. Multiply 2 × 4 = 8, write that eight under the 9, and subtract. This finds us the remainder of 1.

Check: 4 × 402 + 1 = 1,609

6. Practice some more. Subtract to find the remainder in the ones. Check your answer by multiplying the divisor times the quotient, and then adding the remainder. You should get the dividend.

a. $3 \overline{)1\ 2\ 8}$

b. $3 \overline{)9\ 5}$

c. $6 \overline{)4\ 2\ 6\ 7}$

d. $4 \overline{)2\ 8\ 4\ 5}$

e. $5 \overline{)5\ 5\ 0\ 7}$

f. $2 \overline{)8\ 0\ 6\ 3}$

7. Divide these numbers mentally. Remember, you can always check by multiplying!

a. $440 \div 4 =$	b. $3600 \div 400 =$	c. $824 \div 2 =$
$820 \div 2 =$	$369 \div 3 =$	$560 \div 90 =$

Long Division 2

Long division is a process of dividing in parts, starting from the biggest place value unit. For example, we divide the hundreds first, then the tens, then the ones. At each step, if we have a remainder, we combine that with the next unit we are getting ready to divide.

Example 1. Divide 78 by 3.

First we divide the 7 tens by 3. That gives 2 tens for the quotient, and 1 ten left over that we could not divide. The 1 leftover ten is combined with the 8 ones. That is 18. Next, divide 18 by 3. That is 6 and there is no remainder. So, the division is over. The quotient is 2 tens and 6, or 26. Check: $3 \times 26 = 78$.

If you could understand the above example, you will probably have no problems understanding the long division process as it is usually written out in the long division "corner". But if not, do not worry just yet.

In long division, there are **three** processes going on in each step: 1) divide, 2) multiply and subtract to find the remainder, 3) combine the remainder with the next digit from the dividend.

1. Divide.	2. Multiply and subtract.	3. Drop down the next digit.
t o 2 2) 5 8	t o 2 2) 5 8 - 4 1	t o 2 9 2) 5 8 - 4 ↓ 1 8
Two goes into 5 two times, or 5 tens ÷ 2 = 2 whole tens — but there is a remainder!	To find it, multiply $2 \times 2 = 4$, write that 4 under the five, and subtract to find the remainder of 1 ten.	Next, drop down the 8 of the ones next to the leftover 1 ten. You combine the remainder ten with 8 ones, and get 18.

1. Divide.	2. Multiply and subtract.	3. Drop down the next digit.
t o 2 9 2) 5 8 - 4 1 8	t o 2 9 2) 5 8 - 4 1 8 - 1 8 0	t o 2 9 2) 5 8 - 4 1 8 - 1 8 0
Divide 2 into 18. Place 9 into the quotient.	Multiply $9 \times 2 = 18$, write that 18 under the 18, and subtract.	The division is over since there are no more digits in the dividend. The quotient is 29.

1. Divide.	2. Multiply and subtract.	3. Drop down the next digit.

t o
  2
3) 8 4
	t o
	  2
	3) 8 4
	 - 6
	  2
		t o
		 2 8
		3) 8 4
		 - 4 ↓
		  2 4

Three goes into 8 two times, or 8 tens ÷ 3 = 2 whole tens — but there is a remainder!	To find it, multiply 2 × 3 = 6, write that 6 under the eight, and subtract to find the remainder of 2 tens.	Next, drop down the 4 of the ones next to the leftover 2 tens. You combine the remainder tens with 4 ones, and get 24.

1. Divide.	2. Multiply and subtract.	3. Drop down the next digit.

t o
 2 8
3) 8 4
 - 6
  2 4
	t o
	 2 8
	3) 8 4
	 - 6
	  2 4
	 - 2 4
	    0
		t o
		 2 8
		3) 8 4
		 - 6
		  2 4
		 - 2 4
		    0

Divide 3 into 24. Place 8 into the quotient.	Multiply 8 × 3 = 24, write that 24 under the 24, and subtract.	The division is over since there are no more digits in the dividend. The quotient is 28.

1. Divide.

a. 4) 6 4

b. 3) 7 2

c. 3) 8 7

d. 5) 7 5

e. 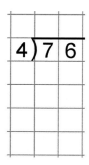 4) 7 6

f. 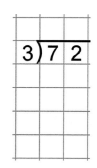 2) 7 8

g. 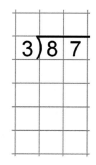 3) 7 8

h. 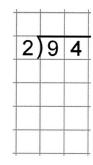 2) 9 4

There are not enough hundreds, so look at two digits in the dividend. You can place a zero in the quotient, or omit it.		

h t o
  0 5 4
4) 2 1 6
  -2 0
     1 6
    -1 6
       0
	h t o
	  0 6 5
	5) 3 2 5
	  -3 0
	     2 5
	    -2 5
	       0
		h t o
		  0 3 8
		7) 2 6 6
		  -2 1
		     5 6
		    -5 6
		       0

2. Divide. Check each answer by multiplying.

a. 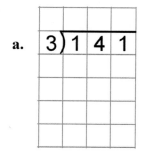 3)1 4 1

b. 4)1 7 2

c. 6)3 8 4

d. 8)2 7 2

e. 3)2 5 2

f. 7)4 0 6

21

Long Division 3

In this lesson we will divide three-digit numbers.

1. Divide.	2. Multiply and subtract.	3. Drop down the next digit.
| ```
  h  t  o
     1
2 ) 2  7  8
``` | ```
  h  t  o
     1
2 ) 2  7  8
  - 2
     0
``` | ```
  h  t  o
     1
2 ) 2  7  8
  - 2 ↓
     0  7
``` |
| Two goes into 2 one time, or 2 hundreds ÷ 2 = 1 hundred. | Multiply 1 × 2 = 2, write that 2 under the two, and subtract to find the remainder of zero. | Next, drop down the 7 of the tens next to the zero. |

1. Divide.	2. Multiply and subtract.	3. Drop down the next digit.
| ```
  h  t  o
     1  3
2 ) 2  7  8
  - 2
     0  7
``` | ```
  h  t  o
     1  3
2 ) 2  7  8
  - 2
     0  7
      - 6
         1
``` | ```
  h  t  o
     1  3
2 ) 2  7  8
  - 2
     0  7
      - 6
         1  8
``` |
| Divide 2 into 7. Place 3 into the quotient. | Multiply 3 × 2 = 6, write that 6 under the 7, and subtract to find the remainder of 1 ten. | Next, drop down the 8 of the ones next to the 1 leftover ten. |

1. Divide.	2. Multiply and subtract.	3. Drop down the next digit.
| ```
  h  t  o
     1  3  9
2 ) 2  7  8
  - 2
     0  7
      - 6
         1  8
``` | ```
  h  t  o
     1  3  9
2 ) 2  7  8
  - 2
     0  7
      - 6
         1  8
       - 1  8
            0
``` | ```
  h  t  o
     1  3  9
2 ) 2  7  8
  - 2
     0  7
      - 6
         1  8
       - 1  8
            0
``` |
| Divide 2 into 18. Place 9 into the quotient. | Multiply 9 × 2 = 18, write that 18 under the 18, and subtract to find the remainder of zero. | There are no more digits to drop down. The quotient is 139. |
```

Can you follow these examples without the explanations?

Dividing the 8 hundreds.	Dividing the 25 tens.	Dividing the 12 ones.	Dividing the 7 hundreds.	Dividing the 17 tens.	Dividing the 21 ones.
1 $6\overline{)852}$ $\underline{-6}$ 2	14 $6\overline{)852}$ $\underline{-6}$ 25 $\underline{-24}$ 1	142 $6\overline{)852}$ $\underline{-6}$ 25 $\underline{-24}$ 12 $\underline{-12}$ 0	2 $3\overline{)771}$ $\underline{-6}$ 1	25 $3\overline{)771}$ $\underline{-6}$ 17 $\underline{-15}$ 2	257 $3\overline{)771}$ $\underline{-6}$ 17 $\underline{-15}$ 21 $\underline{-21}$ 0

1. Divide. Check each division with multiplication. The gridlines will help you keep your hundreds, tens, and ones lined up.

a. $3\overline{)345}$ Check:

b. $5\overline{)615}$ Check:

c. $4\overline{)976}$ Check:

d. $2\overline{)552}$ Check:

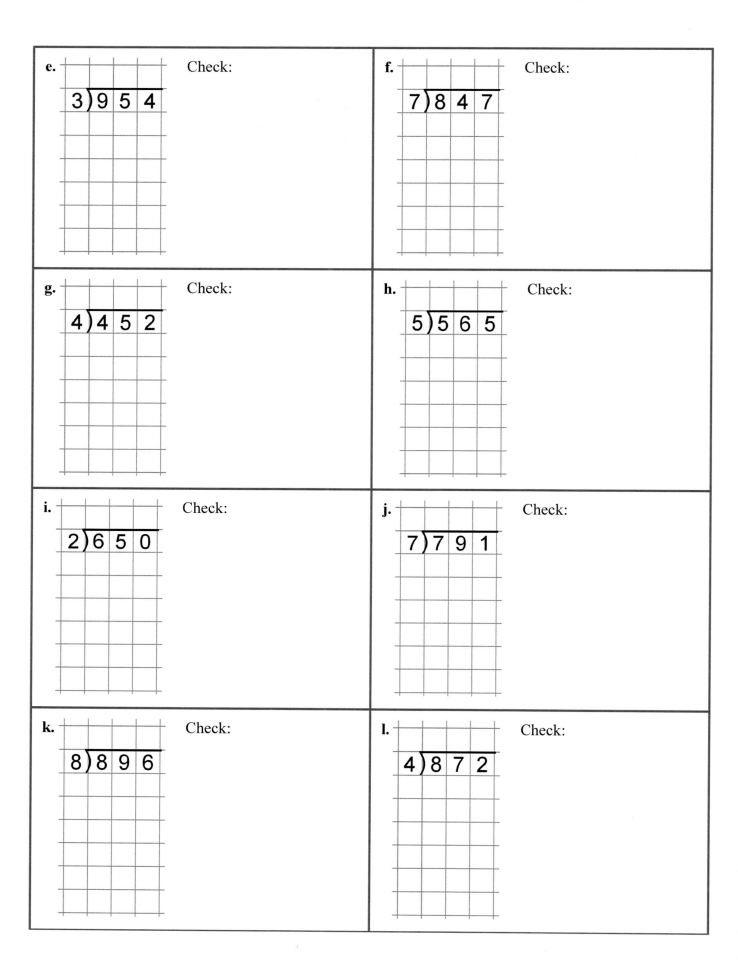

e. 3)954   Check:

f. 7)847   Check:

g. 4)452   Check:

h. 5)565   Check:

i. 2)650   Check:

j. 7)791   Check:

k. 8)896   Check:

l. 4)872   Check:

2. If you need more practice, do these problems as well, using the grids.
   Check each one by multiplying.

a. 567 ÷ 3   Check:

b. 664 ÷ 4   Check:

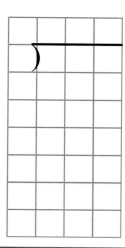

c. 994 ÷ 7   Check:

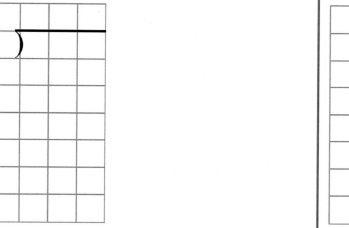

d. 585 ÷ 5   Check:

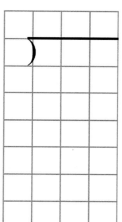

e. 912 ÷ 6   Check:

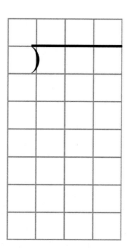

f. 936 ÷ 8   Check:

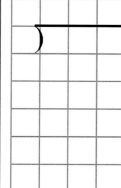

Long Division with 4-Digit Numbers

```
  th h t o          th h t o          th h t o          th h t o
     2                  2 8               2 8 5             2 8 5 9
 2)5 7 1 8          2)5 7 1 8          2)5 7 1 8          2)5 7 1 8
  -4                 -4                 -4                 -4
   1                  1 7                1 7                1 7
                     -1 6               -1 6               -1 6
                        1                1 1                1 1
                                        -1 0               -1 0
                                           1                1 8
                                                           -1 8
                                                              0
```

Check:

```
  2 8 5 9
×       2
_____
```

Long division with 4-digit numbers works the same way as with smaller numbers!

1. Divide. Check each division result with multiplication.

a. 3)7 0 4 1   Check:

b. 4)9 2 4 0   Check:

c. 4)7 1 4 0   Check:

d. 2)9 7 7 0   Check:

2. Divide. Use the grids below. Check each one by multiplication.

a. 5802 ÷ 3   Check:

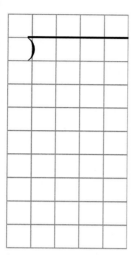

b. 1653 ÷ 3   Check:

c. 9380 ÷ 7   Check:

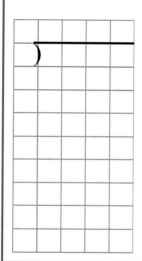

d. 8304 ÷ 8   Check:

e. 7902 ÷ 6   Check:

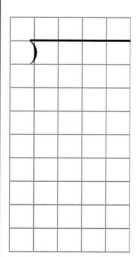

f. 6080 ÷ 5   Check:

	$$\begin{array}{r} 0\ 4 \\ 7\overline{)3\ 0\ 5\ 2} \\ -2\ 8 \\ \hline 2\ 5 \end{array}$$	$$\begin{array}{r} 0\ 4\ 3 \\ 7\overline{)3\ 0\ 5\ 2} \\ -2\ 8 \\ \hline 2\ 5 \\ -2\ 1 \\ \hline 4\ 2 \end{array}$$	$$\begin{array}{r} 0\ 4\ 3\ 6 \\ 7\overline{)3\ 0\ 5\ 2} \\ -2\ 8 \\ \hline 2\ 5 \\ -2\ 1 \\ \hline 4\ 2 \\ -4\ 2 \\ \hline 0 \end{array}$$

There are not enough thousands. So, when you start, look at the **first two digits** of the dividend, and divide the divisor into those.

7 does not go into 3, so look at the **first two digits** ("30"). 7 goes into 30 four times.

7 goes into 25 three times. The remainder is 4 tens.

7 goes into 42 six times.

3. Divide. You may need to look at the first two digits of the dividend. Check your answers.

a. $3\overline{)1\ 4\ 7\ 9}$ Check:

b. $5\overline{)1\ 9\ 2\ 0}$ Check:

c. $6\overline{)5\ 5\ 4\ 4}$ Check:

d. $5\overline{)2\ 4\ 5}$ Check:

e. $6\overline{)5\ 2\ 2}$ Check:

f. $9\overline{)3\ 3\ 3\ 9}$ Check:

4. Two neighbors bought nine trees for $16 each, and shared the cost equally. How much did each person pay?

5. A DVD contains a total of 1,092 minutes of show episodes. The Mayors want to watch it all within a week. They decided to watch the same amount each day.

How many minutes will they watch each day? Express this also as whole hours and minutes.

6. A treasure hunt in the woods that is 2,600 feet long is divided into eight equally long parts, and clues are placed at those points. The children get their first clue at the beginning of the track. At what distance from the start is the second clue?

And the third clue?

7. Cindy bought 8 bags of balloons for her daughter's birthday party, because they were on sale. Each bag had 25 balloons. She took 96 balloons out to give six balloons to each child. How many children were going to be at the party?

How many balloons does Cindy have left?

More Long Division

Study the example carefully. We need to place a **zero** in the quotient.

```
    2 4
4 )9 6 2 0
  -8
   1 6
```

At this point, the division is even. Continue normally, multiplying $4 \times 4 = 16$.

```
    2 4 0
4 )9 6 2 0
  -8
   1 6
  -1 6
     0 2
```

Now, 4 does not go into 02, so **place a zero** in the quotient. You can either continue as usual, OR drop ANOTHER digit from the dividend.

```
    2 4 0 5
4 )9 6 2 0
  -8
   1 6
  -1 6
     0 2 0
      -2 0
         0
```

Dropping another digit from the dividend, we get 020. 4 goes into 20 five times.

1. Let's practice! There will be a zero in the quotient. Multiply to check.

a.
```
5 )5 2 2 5
```
Check:

b.
```
3 )4 2 1 8
```
Check:

c.
```
4 )8 1 4 8
```
Check:

d.
```
7 )9 1 4 9
```
Check:

2. For more practice, do these in a notebook or blank paper. Use grid paper if possible.

a. $8{,}115 \div 3$ **b.** $6{,}540 \div 5$ **c.** $9{,}163 \div 7$ **d.** $6{,}378 \div 6$

Short, even division.	th h t o	th h t o	th h t o

Short, even division.

2,156 ÷ 7 is easy to do mentally:

2,100 ÷ 7 is 300, and 56 ÷ 7 is 8. So the answer is 308.

If we use the division corner, we get a really "short" division. That is because the division is even from the start. So, simply move on to the next digit in the dividend. You do not have to multiply and subtract.

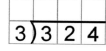

$$\begin{array}{r} 0\ 3 \\ 7\overline{)2\ 1\ 5\ 6} \end{array}$$

7 goes into 21 three times.

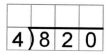

$$\begin{array}{r} 0\ 3\ 0 \\ 7\overline{)2\ 1\ 5\ 6} \end{array}$$

7 goes into 5 zero times...

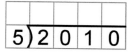

$$\begin{array}{r} 0\ 3\ 0\ 8 \\ 7\overline{)2\ 1\ 5\ 6} \end{array}$$

...so look at the two digits (56). 7 goes into 56 eight times.

3. Divide.

a.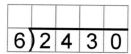

$$3\overline{)3\ 2\ 4}$$

b.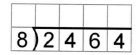

$$4\overline{)8\ 2\ 0}$$

c.

$$5\overline{)2\ 0\ 1\ 0}$$

d.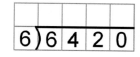

$$6\overline{)2\ 4\ 3\ 0}$$

e.

$$8\overline{)2\ 4\ 6\ 4}$$

f.

$$6\overline{)6\ 4\ 2\ 0}$$

4. Mary divided her 285 buttons evenly into the five compartments. Find out how many buttons are

   a. in one compartment

   b. in three compartments

   c. in four compartments.

5. A gallon of ice cream costs $12.96. You and your brother each pay for 1/8 of the cost, and mom will pay the rest.

   a. Find each person's share of the cost.

     You: _____ Your brother: _____

     Mom: _____

   b. Find each person's share of the ice cream in cups.
     (Hint: 1 gallon is four quarts, and a quart is four cups.)

     You: _____ Your brother: _____

     Mom: _____

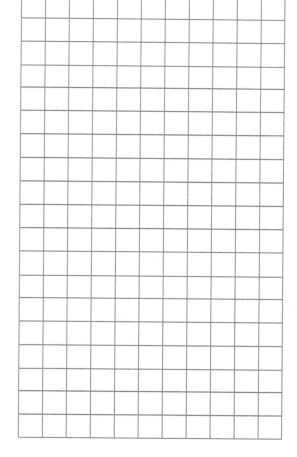

(This page is optional.)

6. Here you can try your division skills with a 5-digit dividend.

   The process is the same, just longer, since the dividend has more digits.

   Lastly, check with multiplication, as always.

a. 3)63702   Check:

b.  2)70814   Check:

c. 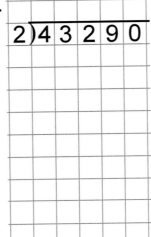 2)43290   Check:

d. 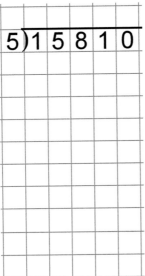 5)15810   Check:

e. 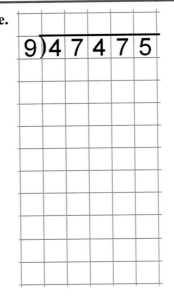 9)47475   Check:

Remainder Problems

When using long division, the division is not always exact.

| At this point there are no more digits to drop down from the dividend. The last subtraction yields 6, which is the remainder. So 125 ÷ 7 = 17, R6.<br><br>Note that the remainder 6 is LESS THAN 7, the divisor. | ```
        1 7
    7)1 2 5
      - 7
        5 5
      - 4 9
          6
``` |

To check:

Multiply the answer (17) by the divisor (7), and then add the remainder (6).

You get the original dividend (125).

```
        4
      1 7
    ×   7
    1 1 9
  +     6
    1 2 5
```

1. Divide. Check each result in the empty space by multiplication and addition.

a. 514 ÷ 3   Check:

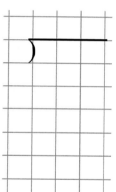

b. 673 ÷ 8   Check:

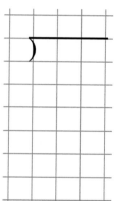

c. 1,905 ÷ 6   Check:

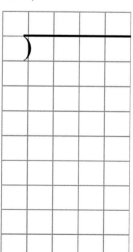

d. 8,205 ÷ 4   Check:

```

2. Find the divisions that are incorrect. Redo the ones that are wrong below.

a.
$$
\begin{array}{r}
7\,7 \\
6\,)\overline{4\,6\,3} \\
-4\,2 \\
\hline
4\,9 \\
-4\,2 \\
\hline
7
\end{array}
$$

b.
$$
\begin{array}{r}
3\,5\,3 \\
7\,)\overline{2\,4\,7\,3} \\
-2\,1 \\
\hline
3\,7 \\
-3\,5 \\
\hline
2\,3 \\
-2\,1 \\
\hline
2
\end{array}
$$

c.
$$
\begin{array}{r}
3\,5\,1 \\
9\,)\overline{4\,0\,5\,9} \\
-3\,6 \\
\hline
4\,5 \\
-4\,5 \\
\hline
0\,9
\end{array}
$$

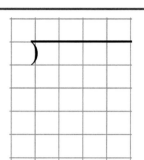

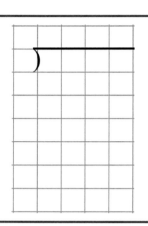

d. How can you spot the error in (a) just by looking at the remainder 463 ÷ 6 = **77 R 7** ?

3. Write a division sentence for each problem, and solve it. Lastly, explain what the answer means.

a. Arrange 112 chairs into rows of 9.

_____

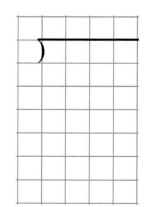

We get _____ rows, _____ chairs in each

row, and _____

_____

b. Arrange 800 erasers into piles of 3.

_____

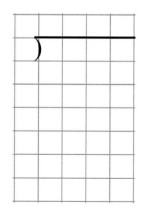

We get _____ piles, _____ erasers in each

pile, and _____

_____

Imagine you are trying to pack some things evenly into some "containers", and they do not go evenly. The last container will not be full!

Often students make mistakes with such problems. Read the question carefully. Sometimes you **DO** need to count the container that is not full, sometimes **not**.

146 people were transported in vans that each carry 9 passengers. How many vans were needed?	You pack 1,250 blank CDs into boxes of 200 each. How many **full** boxes will you get?
The division is $146 \div 9 = 16$ R2, so 16 vans will be full and one van will have 2 passengers. But the *answer* is, they needed **17** vans.	$6 \times 200 = 1,200$; so $1,250 \div 200 = 6$ R50. 6 boxes will be full (and 50 CDs are left over, not packed).

Solve the problems. Sometimes you can use *multiplication* instead of division.

4. A company bags 2,000 lb of potatoes into 12-lb bags.
   The division is: $2,000 \div 12 = 166$ R8.
   How many full bags will they get?

5. If you can fit 50 people into one bus, how many buses would you need to transport 940 people?

6. Mr. Eriksson can have 75 days of vacation each year.
   He wants to divide those days into 4 vacations.
   How long should he make his vacations?
   Make them as close to the same length as possible.

7. A farm packs 400 kg of strawberries so that they make ninety 2 kg boxes, forty 4 kg boxes, and the rest is packed into 6 kg boxes. How many *full* 6 kg boxes will they get?

8. Can you pack 412 tennis balls into containers evenly so that each container has

   a. 4 balls?

   b. 5 balls?

   c. 6 balls?

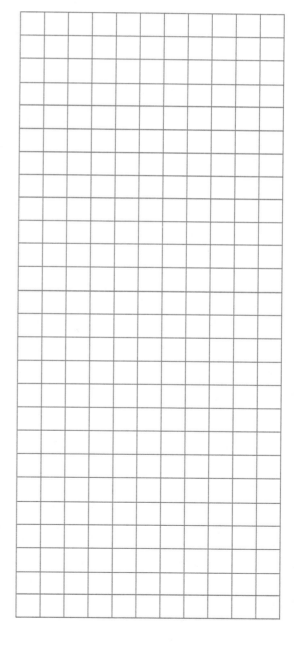

9. Mr. Sandback wants to paint 740 blocks with 6 different colors—red, orange, yellow, green, blue, and purple—in *nearly* equal amounts. How many should be colored with each color?

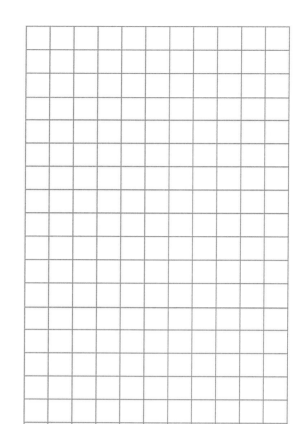

10. Do **one problem** from each box by long division. Can you then figure out the answers to the other two in each box, without actually dividing?

a. $211 \div 3 =$ $212 \div 3 =$ $213 \div 3 =$	**b.** $1{,}206 \div 7 =$ $1{,}207 \div 7 =$ $1{,}208 \div 7 =$
c. $411 \div 5 =$ $412 \div 5 =$ $413 \div 5 =$	**d.** $7{,}185 \div 9 =$ $7{,}186 \div 9 =$ $7{,}187 \div 9 =$

11. *A challenge:* if $231 \div 6 = 38$ R3, then figure out what $232 \div 6$ is.

12. Divide these numbers by 10 and indicate the remainder. There is a shortcut!

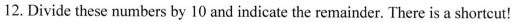

a. $787 \div 10 =$	**b.** $452 \div 10 =$	**c.** $463 \div 10 =$
$66 \div 10 =$	$509 \div 10 =$	$982 \div 10 =$
$340 \div 10 =$	$52 \div 10 =$	$925 \div 10 =$

Puzzle Corner   What is wrong with the division result $31 \div 6 = 4$ R7? After all, $4 \times 6 + 7 = 31$, so it seems to check fine. Explain.

Long Division with Money

Long division with money amounts is done the same way as with whole numbers.
We just place a decimal point in the quotient in the same place as where it is in the dividend.

Complete the divisions below, and check them with multiplication.
After completing the problems 1 a and 1 b, check with your teacher whether you can continue.

1. Divide and check with multiplication.

a. $25.41 ÷ 3   Check:

decimal point
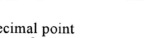

b. $14.88 ÷ 4   Check:

decimal point
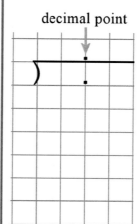

2. Solve the problems.

a. Amy, Sally, and Joe shared equally a salary of $85.50. How much did each one get?

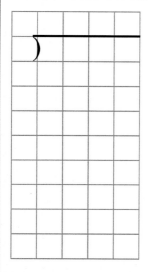

b. If a gallon of milk costs $4.56, what would one quart cost?

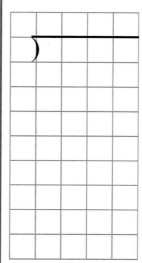

Example. Four people share evenly the cost of a $25.55 pizza. What is each person's share?

The division result is $25.55 ÷ 4 = $6.38 R3¢. The division is not even.

So each person's share is $6.38. But someone has to cover the 3¢, so in reality we could have three persons cover $6.39 and one person covers $6.38.

3. Margie and Annie bought a handbag for $25.56, a mirror for $3.55, and a brush for $2.75. They shared the cost equally. How much did each girl pay?

4. Three people share the cost of movie tickets that cost $25.95 and some popcorn for $4.35. How much was each person's share?

5. A store was selling socks at only 1/4 of the normal price of $3.52. Margie bought three pairs. What was the total cost?

6. Joe bought a $358.60 camera in a store. His dad paid $100 of it for him, and the rest of the payment was divided into four equal monthly payments. How much was each payment?

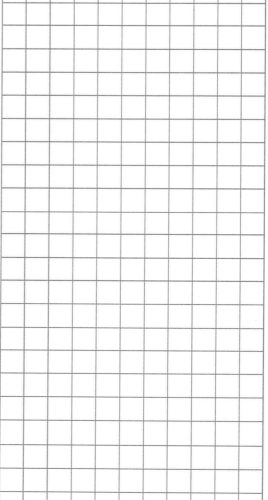

Long Division Number Puzzle

1. Divide. Place each answer in the cross-number puzzle.
   Use your notebook or the grid below.

Across:

a. 3,440 ÷ 8

b. 574 ÷ 7

c. 234 ÷ 9

d. 1,707 ÷ 3

e. 4,756 ÷ 2

Down:

a. 1,072 ÷ 8

b. 6,135 ÷ 3

c. 145 ÷ 5

d. 2,652 ÷ 4

e. 1,442 ÷ 7

f. 3,474 ÷ 9

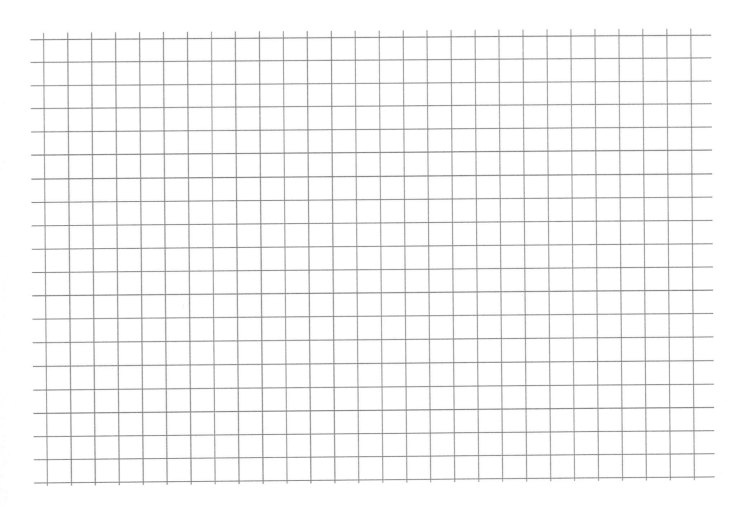

Long Division Answer Key

The Remainder, Part 1, p. 9

1. a. $10 \div 3 = 3$ R1   b. $17 \div 5 = 3$ R2   c. $11 \div 4 = 2$ R3

2. a. $14 \div 3 = 4$ R2   b. $7 \div 2 = 3$ R1   c. $19 \div 3 = 6$ R1
   d. $13 \div 5 = 2$ R3   e. $18 \div 4 = 4$ R2   f. $10 \div 4 = 2$ R2

3.

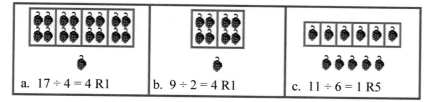

a. $17 \div 4 = 4$ R1   b. $9 \div 2 = 4$ R1   c. $11 \div 6 = 1$ R5

4. a. $10 \div 4 = 2$ R2   b. $17 \div 2 = 8$ R1   c. $12 \div 5 = 2$ R2

5.

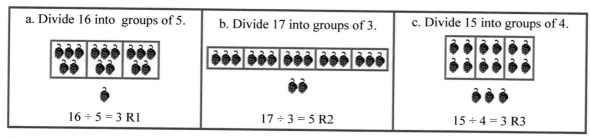

a. Divide 16 into groups of 5.	b. Divide 17 into groups of 3.	c. Divide 15 into groups of 4.
$16 \div 5 = 3$ R1	$17 \div 3 = 5$ R2	$15 \div 4 = 3$ R3

6.

a. $27 \div 5 = 5$ R2 5 goes into 27 five times.	b. $16 \div 6 = 2$ R4 6 goes into 16 two times.	c. $11 \div 2 = 5$ R1 2 goes into 11 five times.
d. $37 \div 5 = 7$ R2	e. $26 \div 3 = 8$ R2	f. $56 \div 9 = 6$ R2
g. $43 \div 5 = 8$ R3	h. $34 \div 6 = 5$ R4	i. $40 \div 7 = 5$ R5

7.

a. $23 \div 4 = 5$ R3 $23 \div 5 = 4$ R3	b. $16 \div 7 = 2$ R2 $20 \div 3 = 6$ R2	c. $21 \div 8 = 2$ R5 $12 \div 9 = 1$ R3

8.

a. $10 \div 5 = 2$ R 0 $11 \div 5 = 2$ R1 $12 \div 5 = 2$ R2 $13 \div 5 = 2$ R3 $14 \div 5 = 2$ R4 $15 \div 5 = 3$ R0	b. $17 \div 3 = 5$ R2 $18 \div 3 = 6$ R0 $19 \div 3 = 6$ R1 $20 \div 3 = 6$ R2 $21 \div 3 = 7$ R0 $22 \div 3 = 7$ R1	c. $12 \div 4 = 3$ R0 $13 \div 4 = 3$ R1 $14 \div 4 = 3$ R2 $15 \div 4 = 3$ R3 $16 \div 4 = 4$ R0 $17 \div 4 = 4$ R1

9. a. $27 \div 5 = 5$ R2   b. $19 \div 5 = 3$ R4
   c. $36 - 3 \div 6 = 5$ R3   d. <u>No</u>, because $51 \div 8 = 6$ R3.
   e. Of four, <u>no</u>. $35 \div 4 = 8$ R3. (The division is not even.)
      Of five, <u>yes</u>. $35 \div 5 = 7$.
      Of six, <u>no</u>. $35 \div 6 = 5$ R5.
      Of seven, <u>yes</u>. $35 \div 7 = 5$.
   f. $38 \div 6 = 6$ R2. There were <u>two photos</u> on the last page. <u>Six pages</u> were full.

1. a. 3   b. 9   c. 7   d. 9

2. a.
$$\begin{array}{r} 6 \\ 5)\overline{3\,2} \\ -3\,0 \\ \hline 2 \end{array}$$
   b.
$$\begin{array}{r} 8 \\ 5)\overline{4\,4} \\ -4\,0 \\ \hline 4 \end{array}$$
   c.
$$\begin{array}{r} 6 \\ 6)\overline{3\,7} \\ -3\,6 \\ \hline 1 \end{array}$$
   d.
$$\begin{array}{r} 4 \\ 7)\overline{2\,9} \\ -2\,8 \\ \hline 1 \end{array}$$

   e.
$$\begin{array}{r} 5 \\ 8)\overline{4\,6} \\ -4\,0 \\ \hline 6 \end{array}$$
   f.
$$\begin{array}{r} 5 \\ 9)\overline{5\,2} \\ -4\,5 \\ \hline 7 \end{array}$$
   g.
$$\begin{array}{r} 8 \\ 4)\overline{3\,5} \\ -3\,2 \\ \hline 3 \end{array}$$
   h.
$$\begin{array}{r} 6 \\ 9)\overline{5\,7} \\ -5\,4 \\ \hline 3 \end{array}$$

3. a. $6 \times 5 + 2 = 32$ b. $8 \times 5 + 4 = 44$ c. $6 \times 6 + 1 = 37$ d. $4 \times 7 + 1 = 29$
   e. $5 \times 8 + 6 = 46$ f. $5 \times 9 + 7 = 52$ g. $8 \times 4 + 3 = 35$ h. $6 \times 9 + 3 = 57$

4. $33 \div 6 = 5$ R3. Jill needed six containers, but only five were full.

5. $100 \div 42 = 2$ R16. They needed three buses to haul the children.

6. There were two teams of seven and one team of six players.

7. $73 \div 20 = 3$ R13. She needed four folders. Three folders were full.

8. $3 \times 23 + 15 = 84$. He had 84 award stickers.

9. $36 \div 11 = 3$ R3. She put three pencils back into the cabinet.

10.

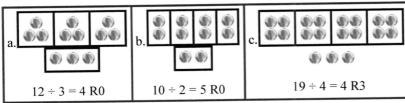

a. $12 \div 3 = 4$ R0	b. $10 \div 2 = 5$ R0	c. $19 \div 4 = 4$ R3

11.

a. $21 \div 5 = 4$ R1	b. $56 \div 8 = 7$ R0	c. $43 \div 7 = 6$ R1
$22 \div 5 = 4$ R2	$57 \div 8 = 7$ R1	$44 \div 7 = 6$ R2
$23 \div 5 = 4$ R3	$58 \div 8 = 7$ R2	$45 \div 7 = 6$ R3
$24 \div 5 = 4$ R4	$59 \div 8 = 7$ R3	$46 \div 7 = 6$ R4

12. The shortcut is: the remainder is always the last digit of the dividend (the number you divide), and the other digits are the quotient (the answer)

a. $29 \div 10 = 2$ R9	b. $78 \div 10 = 7$ R8	c. $54 \div 10 = 5$ R4
$30 \div 10 = 3$ R0	$79 \div 10 = 7$ R9	$55 \div 10 = 5$ R5
$31 \div 10 = 3$ R1	$80 \div 10 = 8$ R0	$56 \div 10 = 5$ R6

Puzzle Corner:
a. $16 \div 5 = 3$ R1 OR $16 \div 3 = 5$ R1
b. $31 \div 7 = 4$ R3 OR $31 \div 4 = 7$ R3
c. $135 \div 4 = 30$ R3 OR $135 \div 30 = 4$ R3

Long Division 1, p. 15

1.

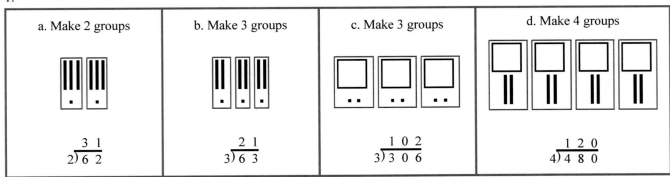

a. Make 2 groups	b. Make 3 groups	c. Make 3 groups	d. Make 4 groups

2. a. 21   b. 131   c. 220   d. 2,010   e. 22   f. 1,006   g. 110   h. 1,201

3. a. 41   b. 71   c. 60   d. 31   e. 92   f. 61   g. 611   h. 601   i. 710   j. 901

4.

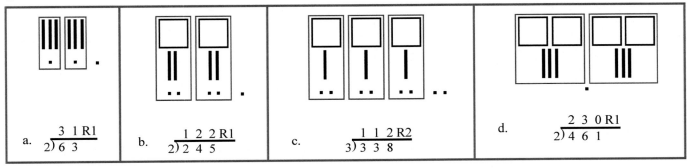

5. a. 211 R3   b. 34 R1   c. 122 R1   d. 22 R1   e. 60 R1   f. 300 R5   g. 30 R5   h. 310 R2

6. a. 42 R2   b. 31 R2   c. 711 R1   d. 711 R1   e. 1,101 R2   f. 4,031 R1

7. a. 110, 410   b. 9, 123   c. 412, 6 R20

Long Division 2, p. 19

1. a. 16   b. 24   c. 29   d. 15   e. 19   f. 39   g. 26   h. 47

2. a. 47   b. 43   c. 64   d. 34   e. 84   f. 58

Long Division 3, p. 22

1. a. 115   b. 123   c. 244   d. 276   e. 318   f. 121   g. 113   h. 113   i. 325   j. 113   k. 112   l. 218

2. a. 189   b. 166   c. 142   d. 117   e. 152   f. 117

Long Division with 4-Digit Numbers, p. 26

1. a. 2,347   b. 2,310   c. 1,785   d. 4,885

2. a. 1,934   b. 551   c. 1,340   d. 1,038   e. 1,317   f. 1,216

3. a. 493   b. 384   c. 924   d. 49   e. 87   f. 371

4. 9 × $16 ÷ 2 = $72. They each paid $72.

5. 1,092 min ÷ 7 = 156 min or 2 h 36 min each day

6. 2600 ÷ 8 = 325. The second clue is at 325 feet. The third clue is at 650 feet.

7. 96 ÷ 6 = 16. Sixteen children were coming to the party. 8 × 25 = 200 and 200 − 96 = 104. She had 104 balloons left.

More Long Division, p. 30

1. a. 1,045   b. 1,406   c. 2,037   d. 1,307

2. a. 2,705   b. 1,308   c. 1,309   d. 1,063

3. a. 108   b. 205   c. 402   d. 405   e. 308   f. 1,070

4. a. 285 ÷ 5 = 57.
   There are 57 buttons in one compartment.
   b. 3 × 57 = 171.
   There are 171 buttons in three compartments.

4. c. 4 × 57 = 228.
   There are 228 buttons in four compartments.

5. a. $12.96 ÷ 8 = $1.62.
   You will pay $1.62. Your brother will pay $1.62.
   Mom will pay $12.96 − $1.62 − $1.62 = $9.72.
   b. You get two cups. Your brother gets two cups.
   Mom gets 12 cups.

6. a. 21,234   b. 35,407   c. 21,645   d. 3,162   e. 5,275

Remainder Problems, p. 33

1. a. 171 R1   Check: 3 × 171 + 1 = 514
   b. 84 R1   Check: 8 × 84 + 1 = 673
   c. 317 R3   Check: 6 × 317 + 3 = 1,905
   d. 2,051 R1   Check: 4 × 2,051 + 1 = 8,205

2. a. wrong; 77 R1   b. right   c. wrong: 451
   d. The remainder is larger than the divisor.

3. a. 112 ÷ 9 = 12 R4. We get 12 rows, 9 chairs each row,
   and 4 chairs will be left over or put in an extra row.
   b. 800 ÷ 3 = 266 R2. We get 266 piles, 3 erasers in each
   pile, and 2 erasers left over.

4. They will get 166 full sacks.

5. 20 × 50 = 1,000 and 19 × 50 = 950. So, 19 buses is
   enough to transport 950 people.

6. 75 ÷ 4 = 18 R3. One 18-day vacation and three 19-day
   vacations. If the division had been even, all of the
   vacations would have been 18 days, but now there are
   three extra days to be added to three of the vacations.

7. 400 − (2 × 90) − (4 × 40) = 60; 60 ÷ 6 = 10.
   They will have ten full 6-kg boxes of strawberries.

8. a. Yes. There will be 103 containers. 412 ÷ 4 = 103.
   b. No, there will be 82 containers with 2 left over.
   412 ÷ 5 = 82 R2.
   c. No, there will be 68 with four left over.
   412 ÷ 6 = 68 R4.

9. 740 ÷ 6 = 123 R2. Paint 123 blocks in four of the colors
   (any four), and 124 blocks in the two remaining colors.

10. a. 70 R1; 70 R2; 70 R3   b. 172 R2; 172 R3; 172 R4
    c. 82 R1; 82 R2; 82 R3   d. 798 R3   798 R4   798 R5
    You can figure out the two other problems after solving
    one, because the remainder will increase by one as
    the dividend increases by one.

11. It would be 38 R4. The only difference is that the
    remainder increases by 1.

12. a. 78 R7; 6 R6; 34   b. 45 R2; 50 R9; 5 R2
    c. 46 R3; 98 R2; 92 R5
    The ones digit of the dividend will always be the
    remainder.

Puzzle Corner: The remainder is larger than the divisor.

Long Division with Money, p. 37

1. a. $8.47   b. $3.72

2. a. $28.50   b. $1.14

3. $25.56 + $3.55 + $2.75 = $31.86
   $31.86 ÷ 2 = $15.93. Each girl paid $15.93.

4. ($25.95 + $4.73) ÷ 3 = $10.10. Each person's share
   was $10.10

5. The cost was $3.52 ÷ 4 × 3 = $2.64.

6. $358.60 − $100 = $258.60; $258.60 ÷ 4 = $64.65.
   Each payment was $64.65.

Long Division Number Puzzle, p. 39

1. Across:
   a. $3{,}440 \div 8 = 430$
   b. $574 \div 7 = 82$
   c. $234 \div 9 = 26$
   d. $1{,}707 \div 3 = 569$
   e. $4{,}756 \div 2 = 2{,}378$

Down:
   a. $1{,}072 \div 8 = 134$
   b. $6{,}135 \div 3 = 2{,}045$
   c. $145 \div 5 = 29$
   d. $2{,}652 \div 4 = 663$
   e. $1{,}442 \div 7 = 206$
   f. $3{,}474 \div 9 = 386$

a. 1					
3		b. 2		b. 8	e. 2
a. 4	3	0			0
		4		c. 2	6
		d. 5	d. 6	9	
			6		f. 3
		e. 2	3	7	8
					6

Appendix: Common Core Alignment

The table below lists each lesson, and next to it the relevant Common Core Standard.

Lesson	page number	Standards
The Remainder, Part 1	9	4.OA.3
The Remainder, Part 2	12	4.OA.3
Long Division 1	15	4.NBT.6
Long Division 2	19	4.NBT.6
Long Division 3	22	4.NBT.6
Long Division with 4-Digit Numbers	26	4.OA.3 4.NBT.6
More Long Division	30	4.OA.3 4.NBT.6
Remainder Problems	33	4.OA.2 4.OA.3 4.NBT.6
Long Division with Money	37	4.MD.2
Long Division Number Puzzle	39	4.NBT.6

Made in the USA
Middletown, DE
28 July 2017